What Are They Good For?

by Barbara Wood

A bison has big horns.
What are they good for?

Fighting!

A prairie chicken has brown and tan feathers. What are they good for?

Hiding!

An antelope has long, strong legs.
What are they good for?

Running!

A hawk has sharp claws.
What are they good for?

Hunting!

A jackrabbit has long back feet.
What are they good for?

Hopping!

Special parts help prairie animals survive in their habitat.

A bison can use its horns to fight.

A prairie chicken can hide in the grasses of the prairie.

An antelope can run with its long, strong legs.

A hawk can use its sharp claws to hunt for food.

A jackrabbit can hop away from animals that chase it.